The Brain Book for Kids

Discovering the Mysteries Inside Your Head

By Taylor Ellyn Butler

Illustrated by Riki Artana

First Edition

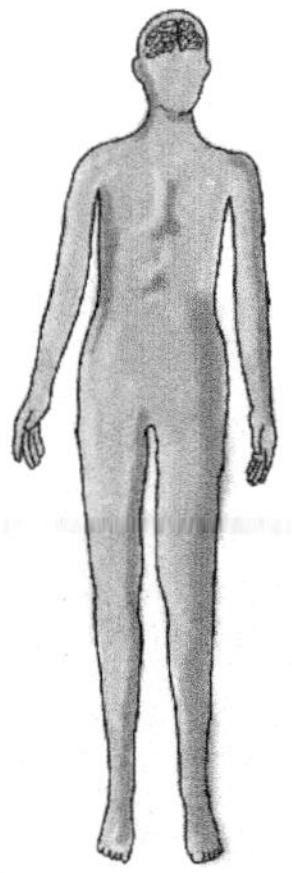

Chapter 1: Meet Your Mind

What is the brain and what does it do?

Who are you?

What makes you, you?

You might answer with your name, or where you're from, or maybe by describing some of your favorite things. But how did those preferences form?

How are you able to understand and answer the question in the first place?

The truth is, a lot goes in to making you! From the genetic blueprint provided by your parents, to the food you eat, the language you speak, and the experiences you have, these all combine to create the unique person that you are. And the team captain of all these players? You guessed it: your brain!

All the organs, bones, muscles, and parts of your body have a function. At the center of them all is the operating system, the conductor of the orchestra, the main computer: your brain. It gathers information from the rest of your body and from the world around you, it interprets and analyzes and consolidates that information and sends signals to your body to move, talk, play, sing, and do all the things that make you, you!

CONNECTIONS

Your brain is all about making connections. As part of the Central Nervous System, your brain teams up with your Spinal Cord to connect its computing power to the rest of your body, otherwise known as the Peripheral Nervous System. From your beating heart, to your expanding lungs, to each finger and toe, every part of your body is connected to your brain by little fibers called Nerves. They extend out from your Spinal Cord, carrying signals from your brain telling your arm to reach out for that chocolate bar, or carrying signals back from your toe that something sharp is on the ground… ouch!

These connections between your Central and Peripheral Nervous Systems help keep your body balanced. Scientists call this relationship homeostasis: maintaining a stable internal environment with an unstable external environment. So, on a hot, sunny day, your brain tells your skin to sweat in order to cool itself down. Or if you're caught out in the rain without a coat, your brain will tell your muscles to shiver, generating heat within the body.

These feedback loops between your brain and body help regulate all kinds of things, from your bladder and thirst, to breathing and heart rate, to hormone levels and blood pressure. And your brain does all this without you even having to think about it! Your brain is constantly hard at work keeping your body healthy and homeostatic.

Your brain is also connected to the outside world through your five senses. Sight, Touch, Taste, Smell & Hearing are how we gather information about the world around us, and our brain interprets, integrates, and responds to that information.

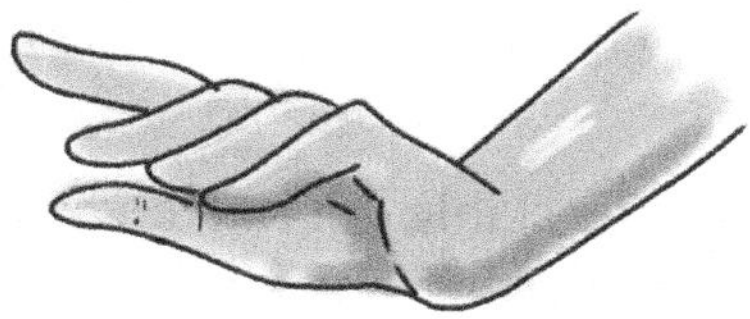

Your brain even makes connections with itself! Making new connections within the brain is essentially how we learn. From learning the basics like walking, we're forming connections between the parts of the brain that deal with sight with movement. Add in connections to areas of the brain that interpret sound, and you might be learning to play an instrument, or speak a new language. Some of these new connections become stable, creating memories, and influence how we make decisions or interpret new information.

You may have heard the phrase before, "everything is connected." Well, as far as your brain is concerned, it's true! The connections between your brain and your body, the outside world, and even within itself, are how we are able to do just about anything. We'll dive into the many wonderful things our brains are capable of because of these connections. First, let's take a closer look at the brain itself. Let's find out what exactly these connections are made of.

COMPONENTS

THE NEURON

Key Terms

Central Nervous System: the brain and spinal cord, the brain acting as the main operating system and the spinal cord connecting the brain to the Peripheral Nervous System.

Peripheral Nervous System: nerves and other components of the nervous system that reside outside the brain and spinal cord, connecting the entire body to the Central Nervous System.

Nerves: Long fibers extending out from the spinal cord to the body, like wires connecting organs and muscles to the Central Nervous System.

Homeostasis: maintaining a stable internal environment with an unstable external environment.

Your entire body is made up of tiny cells, which are the building blocks of life. But some cells look different than others, and your brain is made up of a special kind of cell called a Neuron. These brain cells, or neurons, are special because they have long tails, called Axons, that can extend out, grow, and make those oh-so-important connections with other neurons. We call the connections between neurons Synapses, which are actually little pockets of space. So how do neurons talk to each other if they aren't actually touching? We need tiny chemical messengers, called neurotransmitters, to travel across the Synapse from one neuron to the next.

This is one way that neurons can have vastly complex communication networks, because our brains are filled with many different kinds of neurotransmitters, which specialize in certain functions. For example, similar to how you might have different interests or talents than your friend, neurotransmitters do different jobs in the brain. Maybe you are really good at art, whereas your friend is a really good soccer player.

For our brain's chemical messengers, dopamine is a key player in motivation, whereas serotonin helps you sleep at night. Scientists are still discovering new neurotransmitters and uncovering the secrets of their specialties, just like you continue to discover new talents and interests, too.

Key Terms

Neuron: a specialized brain cell.

Axon: the long "tail" of a neuron.

Synapse: the connection point between two neurons, which is actually a tiny, empty space between the end of an axon and another neuron.

Neurotransmitter: little chemical messengers that travel the space of the synapse to allow communication between two neurons.

Glial cells: brain cells that act as "helpers" to chatty neurons, providing support, nutrients, or insulation for faster communication.

Neurons are also accompanied by Glial cells, or helper cells, that help neurons in lots of different ways. Some provide structural support, others provide nutrients and help maintain the health of neurons, while others help speed up the communication between neurons. These helper cells are also still revealing their special talents to Scientists today.

THE LOBES

With these building blocks, your skull is packed full of these talkative neurons, constantly chatting away to each other and to your body. We'll discuss exactly how neurons communicate in Chapter 3, but for now let's look at a broader picture.

Your brain can be divided into several areas, specializing whole groups of neurons to do different jobs. We call these specialized areas the Lobes of the brain, where neurons take on specific jobs, similar to how there are different roles or characters in any classic adventure story. You've got to have your leader and strategist, guiding the group towards a fantastic goal, like a wise wizard hoping to destroy an evil ring in the mountains of Mordor. You have the explorer, always taking note of their surroundings and watching out for danger, like a ranger from the North. You have the emotional one always telling stories or making friends, and of course your artsy and creative friend to liven up the group. Each person plays an important part in the overall story, just like each lobe plays an important role in your brain.

The Frontal Lobe is like the leader or strategist of the group. Responsible for decision-making, planning, reasoning, and problem-solving, your leader is often at the front and center of the group just like your Frontal Lobe sits in the front and center of your brain. This region of neurons has an ability called Executive Function. This allows them to manage and regulate cognitive processes. This is like how a good leader delegates and uses the strengths of their friends. Sometimes leaders can be impulsive, especially young leaders like Harry Potter or Luke Skywalker. This makes sense, because your Frontal Lobe isn't fully developed until around age 25!

Key Terms

Executive Function: the brain's ability to manage and regulate various forms of "thinking," such as planning, organizing, paying attention, decision-making, etc.

Neural Plasticity: the brain's ability to change and adapt.

Sitting just behind the Frontal Lobe is the Parietal Lobe, which is like the explorer of the group. Always scoping out the area, finding clues or a path through the forest, or the first to hear that "strange sound in the distance." The Parietal Lobe is all about sensory information, using spatial awareness and perception to help you navigate the world.

On either side of your brain are your Temporal Lobes, which are like storytellers of the group. They might be a little emotional, as your Temporal Lobes deal with regulating emotions, as well as forming memories, understanding language, and integrating information. But these are all the components of a great storyteller, using memories and character voices to reveal emotions through an exciting journey.

Finally, this adventure group needs some color, in the artsy and stylish visionary of the group: The Occipital Lobe. Located at the back of your brain, the Occipital Lobe processes visual information. It allows you to process color, movement, shapes, texture, and objects. Just as the visionary of the group has the biggest imagination, seeing beauty, wonder, and possibilities where the others can't, the Occipital Lobe is also responsible for vivid imagery.

With these four specialized lobes of the brain using their talents in unique ways, your brain is like a heroic team of adventurers fulfilling a quest every day. Because while they may have different strengths, their true strength lies in their connections, and ability to work as a team to conquer any foe.

FUNCTIONS

You are who you are, you do what you do, see what you see, because of your brain. The rest of this book goes into detail about all of the mind-boggling wonders your brain does every day, which all boil down to connections between neurons. From seeing and hearing, to memory and personality, our brains are each like an expansive universe waiting to be explored.

With the power of 86 billion neurons - close to the number of galaxies in the entire universe - your brain contains over 100 trillion connections! That's a whole lot of possibility… and the coolest part? Your brain is always changing. When we learn new things, adapt to changes, grow as a person, our brain is making new connections and pathways every day. Scientists call this flexibility of the brain Neural Plasticity, which is just neurons making new connections with each other to create a brain that is uniquely yours.

Did you know? There are about 86 billion neurons in your brain, almost as many stars as in the Milky Way galaxy.

Chapter 2: The Scientific Method

How do we know what we know?

Scientists study many things, from the laws of gravity to dinosaur fossils. The universe is full of mysteries waiting to be uncovered! Whether you are unraveling the intricacies of the earth's environment or deciphering the coded messages between neurons in the brain, scientists all use the same strategy to answer the wide variety of curious questions we have.

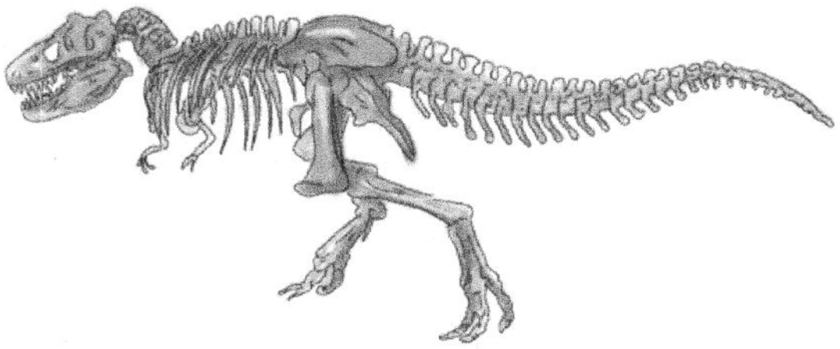

CURIOUS QUESTIONS

Every scientist begins with a question. What makes the planets move? What causes global warming? How does the brain remember different faces? Your question may be a broad one, like how does memory work? Or it might be more specific, like what is the function of serotonin in the sleep-wake cycle? No question is too big or too small for the curious scientist.

What are you curious about? Are there any unsolved mysteries you would like to understand? Curiosity is the seed of science, sparking grand adventures into the universe. So, how do scientists go about answering these curious questions? Let's journey through the Scientific Method.

GATHER INTEL

Once you have your question, the next step is to gather intelligence on what we already know. Like a detective solving a mystery, scientists gather clues that lead them closer to answering their curious questions. Maybe someone else was also curious about the same thing. Maybe, hundreds of people over many years have been curious about the same thing! We don't always have to start at square one. In fact, one of the best advantages of scientists all using this same process to answer questions is that we can build off one another, continuing the journey decade after decade.

Say you are curious about how the brain processes color. We need to understand all the parts before we can put them all together, so you gather intel on what is color? Color is different wavelengths of light. Ok, how does light get into the brain? Through the eyes, of course. But then how do the eyes translate light into a signal the brain understands? And how can the eyes tell the difference between green and blue? We'll get into all those details in Chapter 4, but no matter your question, other scientists have likely had similar, if not the same questions. Before we sit down to make a prediction, we want to be well-informed about the possibilities, parts, and processes surrounding our question.

YOUR HANDY HYPOTHESIS

You've done your searching, reading everything there is to know about your question, but still there is something unknown, still a mystery to be solved. It's time to form your handy hypothesis! A hypothesis is a very specific kind of statement scientists make, predicting the answer to their question based on all the intel they've gathered. This specific prediction provides the framework for answering your question, like the scaffolding and blueprints used to create a building.

What makes a hypothesis so handy? A prediction that is testable and falsifiable! A good hypothesis, or a good prediction, is one that can be tested and proven wrong. That means you make a prediction that you can test through observation, experimentation, or any means of gathering data. And importantly, that data has the possibility of proving your prediction wrong.

TEST IT OUT

Get your lab coats and safety goggles on, it's experimenting time! Scientists test out their hypotheses in many exciting ways, collaborating with doctors or engineers to create new ways of gathering data. To study the brain, scientists use scanners to look inside the brain or record its activity, observe people in specific tasks or scenarios, develop tests and surveys, and much more. The methods a scientist uses often depend on their question.

AMAZING ANALYTICS

Now that we've gathered data, made our observations, scanned some brains, or completed our tests, we need to make sense of the information. Scientists often use Statistics to organize, analyze, and interpret the data they gather. Statistics are especially important because just like you are a completely unique and complicated individual, the answer to scientific questions can also be unique and complicated, and not as easy as 2+2=4. Statistics allow us to generalize, drawing insights and trends from vastly complex ranges of data. Statistics also allow us to make new predictions with more confidence, keeping the cycle of the scientific method going.

SHARING IS CARING

Scientists share their discoveries in peer-reviewed journals, so that others can learn from them and form new questions and predictions. The peer-review process is important in science because it keeps scientists accountable, making sure they are following the process of the scientific method, and affirms published findings have been assessed for errors, bias, revisions, clarifications, and relevance.

Isaac Newton once said to a friend, "If I have seen further, it is by standing on the shoulders of Giants." He meant that if he was a successful scientist, answering some of the biggest questions of the universe, he was only able to do so because of all the knowledge that was discovered before him. That is the beauty of science, it is always growing. Science builds on the mountains of discoveries, insights, questions, and solved mysteries that came before. That is why it is so important for scientists to share their discoveries, placing one more stone atop the mountain for others to stand on.

Key Terms

Scientific Method: a continual process of prediction and observation that scientists use to build knowledge and understanding of the universe.

Hypothesis: a specific prediction that can be tested and proven wrong.

Theory: the culmination of many iterations of the scientific method, a theory is a well-supported explanation of a phenomenon.

Falsifiability: able to be proven false.

Statistics: branch of mathematics used to organize, analyze, interpret, summarize, and make predictions about data.

Peer Review: a process of revision and anonymous feedback every scientific work goes through before publication.

Did you know? Your brain can generate about 23 watts of power—enough to power a small light bulb!

Chapter 3: The Language of the Brain

Action Potentials

Some people grow up speaking English. Some people speak Spanish, or French, or Mandarin. Some people speak sign language, communicating without sound; others speak brail, communicating without sight. Still others find math to be the most sensible language, or music, communicating through melody or emotion. The brain also has a language of its own, communicating the same way in every living organism. This language is called Action Potentials.

An Action Potential is an electro-chemical signal neurons use to communicate. The "electro" or electric part of this signal means that an Action Potential happens because of a change in the electricity or charge of the neuron. That's why we can measure the brain's "electrical activity," which is just measuring how much neurons are talking to each other. The "chemical" part of this signal means that Action Potentials cause the release of chemicals, otherwise known as neurotransmitters. These little chemical messenger travel from one neuron, across the synapse, to be received by another neuron. This combination of an electrical signal causing the release of chemical messengers allows the brain to communicate in simple, yet complex ways. The language of the brain is both a simple binary signal, where a neuron has a surge of electricity, and at the same time a complex and diverse message, with different neurotransmitters having numerous functions and roles.

SET THE STAGE

Let's get to know the players that make these Action Potentials happen. Just like any language has vocabulary and grammar rules, Action Potentials have key players and rules for how and when to "speak." The conditions need to be just right for an Action Potential to occur, so let's set the stage for neuronal communication.

The neuron is where it all happens, like the stage itself. Remember that a neuron is a brain cell, and like any cell is has a nucleus, cell wall, and all kinds of molecules with specialized functions inside. You can think of all these parts contributing to the whole, just like the stage of a play has a curtain, a backstage, props, and people running about with different roles, like lead actors, dancers, and stage managers. In the neuron, some of these molecules running about are called ions, which are electrically charged, meaning they have a positive or negative charge. Ions can be thought of as actors in the play, and in any good story, you need to have both heroes and villains. In the case of the neuron, a negative charge doesn't mean those ions are "bad," nor a positive charge as "good," but they play different roles while working together to tell a story. Lots of these ions float around inside and outside the cell, passing through little channels, or openings, within the cell wall. Depending on how many are inside or outside the cell, the neuron itself will have a positive or negative charge overall. These ions are important for "setting the stage" for an Action Potential to happen, and their movement, like actors in a play, create the magic of communication.

Key Terms

Action Potential: the electrochemical signal neurons use to communicate. It is a spike in electrical activity that causes the release of neurotransmitters into the synapse.

Electro-chemical signal: a form of communication that involves both changes in electrical charge as well as chemical messengers.

Electrical charge: a property of an atom, molecule, or cell (such as a neuron), can be either positive, neutral, or negative.

Ion: an electrically charged particle, atom, or molecule.

Ion Channel: a "tunnel" in the membrane of a neuron that ions can move through.

THE LOW DOWN

So how exactly does an Action Potential happen? What rules or conditions allow the signal to "fire"? It all comes down to the electrical charge of the neuron. As ions flow in and out of the neuron, the electrical charge of the neuron changes. If positive ions flow into the neuron, the neuron becomes more positive. If positive ions flow out of the neuron, the neuron becomes more negative. And the opposite is true for negative ions flowing in and out of the cell. As these ions dance around, flowing through little channels in the cell wall, they slightly change the overall electrical charge of the neuron.

At some point, these small changes in charge will reach a threshold. A threshold is a special level of electrical charge, that sparks the firing of an action potential like flipping a switch to turn on a light, or pushing the first domino and watching all the rest topple in a chain reaction. This threshold is like the magic button that, once pushed, wham! Off goes the Action Potential! The neuron has a huge surge of electrical activity, which travels wicked fast down the axon, the tail of the neuron, and causes the release of neurotransmitters into the synapse.

Key Terms

Threshold: a limit, or point at which the switch turns on.

Receptors: what neurotransmitters attach to on the receiving neuron across the synapse, which "read" the messages of the neurotransmitters.

INTERPRETING THE SIGNAL

The Action Potential is essentially a surge in electrical activity, where the electrical charge of the neuron changes drastically. Like electricity flows through a wire, this electrical signal travels down the neuron's axon. Once this surge of electricity reaches the end of the axon, it causes the release of neurotransmitters into the synapse. These little chemical messenger travel across the tiny gap between one neuron and the next, to be received by "receptors" on the other side that will "read" their message.

Neurotransmitters can carry all sorts of messages for neurons. Some neurotransmitters are like volume controllers, where some tell the neuron to speak "louder" and others are saying "shhhh." You can activate these kinds of messengers yourself! Have you ever tried to find a friend in a crowd? Maybe they told you they are wearing a red shirt, so you scan the crowd looking for red. While you're doing this, your neurotransmitters are helping you by saying "louder" to your red-detecting neurons and "shhhh" to neurons that detect other colors. When you do finally see your friend, it almost feels like your brain sparks! Because those red-detectors were primed to scream! A similar process happens when you play *Where's Waldo*? Try it for yourself by looking around the room for something that's your favorite color!

Other neurotransmitters have message like "way to go!" providing an internal reward for your brain. Some say "danger!" and warn you about potential hazards while also activating your body to respond. Others still have all kinds of moody messages, telling you to be sad or happy or mad. And there are many more neurotransmitters that do all sorts of things like help you move and dance, remind your belly to be hungry, help you sleep, talk, and see, and all the little tasks that help you navigate the world every day.

Your brain has over 100 different kinds of neurotransmitters, carrying all kinds of different messages. Scientists are still discovering new ones within the brain, as well as trying to understand the messages and complexities of the ones we know about. The language of the brain is beautiful and complex, like a symphony happening within our minds every second. And yet, neurons can achieve this complexity of language with a simple electrical signal, like computers creating digital wonders with a simple binary code. Will you become fluent in the language of the brain, and possibly discover a new kind of message?

Did you know? Information in your brain travels at about 260 miles per hour—that's as fast as a race car!

Chapter 4: Sensation & Perception

Our window to the world

Neurotransmitters and Their Messages

Glutamate: "louder!" The brain's main excitatory neurotransmitter, telling the next neuron to talk more. In other words, it makes the neuron more likely to fire an Action Potential.

GABA: "shhhh!" The brain's main inhibitory neurotransmitter, telling the next neuron to quiet down and talk less. In other words, it makes the neuron less likely to fire an Action Potential.

Dopamine: "way to go!" The brain's reward system, involved in motivation and feel-good behaviors.

Norepinephrine and Epinephrine: "danger!" The brain's alarm system, warning you of potential hazards and activating bodily systems to respond, such as increasing heart rate, boosting energy, and increasing focus. Involved in your body's "fight or flight" response and release of Adrenaline (aka epinephrine).

Serotonin: The brain's main mood regulator, balancing emotions, as well as regulating your sleep-wake cycle.

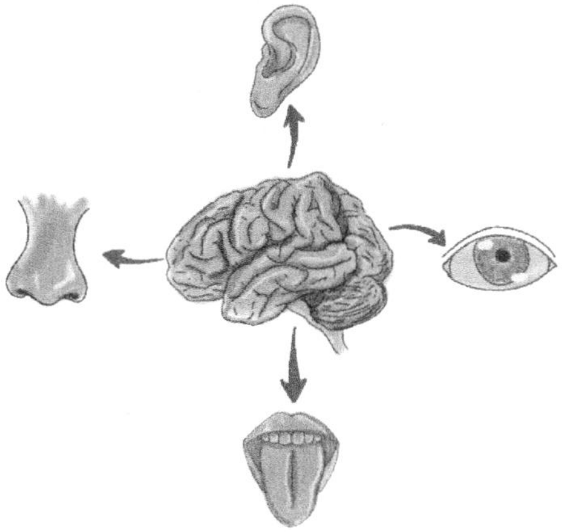

Sensation and Perception are how we navigate the world around us. These are two separate, but related processes. Sensation is the first half of the journey, gathering the information from the outside world. We have five senses: sight, hearing, touch, taste, and smell. We "sense" these things as our bodies interact with the world, like light entering the eye or food on the tongue. But the second half of the process happens in the brain. Perception is how we understand that information, interpreting and making sense of the various signals about color or saltiness.

Sensation is the raw sensory input that informs you of your surroundings, such as feeling the heat of the sun, hearing the rustle of leaves, or seeing the squirrel hop from tree to tree. Our bodies are filled with different kinds of specialized cells that translate information about sight, sound, touch, taste, and smell into the language of the brain - aka action potentials! They send those signals to the brain to make sense of the world around us.

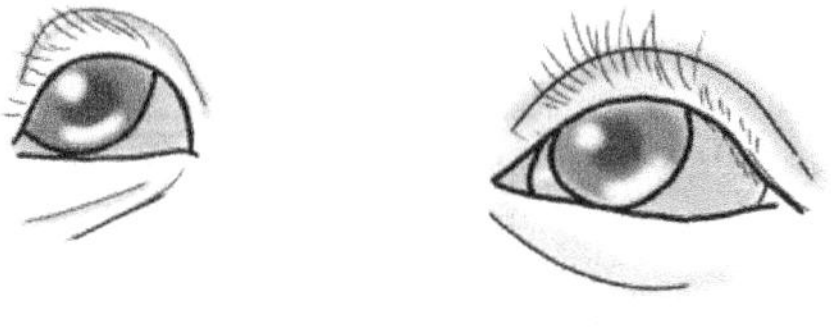

Perception happens in the brain, processing those signals from our senses. Our brain is responsible for making sense of the sensory input to recognize objects, detect patterns, remember people and places. Perception is partly how we develop different preferences and tastes, because it's your brain that decides green is better than blue, or sugar is better than salt. Your eyeball and your tongue are just the windows, the translators, and have no preferences on their own. Our brains integrate all of this incoming information from the senses as well as with the information we have stored in our memories, creating an interactive, movie-like world for us to navigate more easily.

Now let's dive into how our bodies and brains understand these senses:

Key Terms

Sensation: the raw sensory input that informs you of your surroundings, including sight, hearing, touch, taste, and smell.

Perception: the processing of sensory input in order to understand incoming information, recognize objects, detect patterns, remember things, etc.

SIGHT

Sight is all about light! Light is how we are able to see anything, and why we are lost in the dark. The light we can see is what we know as "visible light," but this is actually only a tiny sliver of all the light that exists in the universe, known as the whole electromagnetic spectrum. The electromagnetic spectrum contains all kinds of light that exist in the universe, which have different shapes and properties (aka wavelengths and frequencies). Visible color, the light we can see, is a small part of this spectrum of energy.

Imagine a rainbow, but all you can see is a tiny sliver of green, and the other colors are invisible to you. Red, orange, yellow, blue, and violet are all still there, but you can't see them. That's what the electromagnetic spectrum is like, the rainbow is just that tiny slice of green - the colors we can see are just a small part of the whole. But those other forms of light aren't invisible to everyone - bees, butterflies, and some birds can see in ultraviolet (the form of light just above our detection), and snakes and mosquitos can see in infrared (the form of light just below our detection). Humans have even made devices to detect these other forms of energy for us - like radio waves and x-ray scanners.

Our human eyes are made for detecting visible light. But how? Imagine your eyes are like cameras, constantly filming the world around you. When you look at something, like your favorite toy or a beautiful sunset, light from that object, or light bouncing off that object, enters your eyes. Now, on the back of your eyeballs, there's something called the retina. It's like the film in a camera, or like a screen your movie of the world gets projected onto. The retina is made up of another kind of specialized cell called photoreceptors, which have the special ability of detecting light!

There are two main types of photoreceptors: rods and cones. Rods help you see in low light, like when it's dark outside or in a dimly lit room. But they don't see different colors, which is why everything looks sort of grey in the dark. On the other hand, cones help you see colors, and they are better at seeing details in brighter light.

When light hits these photoreceptor cells, the light activates action potentials in the photoreceptors. Remember that neurons "speak" in action potentials, so these photoreceptors are kind of like light-translators, converting light into a signal that neurons in the brain can understand. These signals get sent from the photoreceptors in the eye to neurons in the brain through a pathway called the optic nerve. Think of the optic nerve as a super-fast highway that carries messages from your eyes to your brain.

Now, your brain is like a supercomputer that processes all these signals it gets from your eyes, mostly in a region called the visual cortex. Some neurons get to work at understanding motion, others figure out shapes, color, objects, or faces. All these neurons work together, making sense of all the information it gets from the different photoreceptors to create the images you see. It's like putting together puzzle pieces to make a complete

Key Terms

Visible light: the small sliver of the electromagnetic spectrum that humans can detect.

Electromagnetic spectrum: the entire range of radiant energy that exists in the universe.

Retina: a wall of cells on the back of your eyeball that light from the outside world is projected onto.

Photoreceptors: specialized cells in your retina that convert light into neural signals - aka action potentials - to send signals to your brain.

Rods: photoreceptors that help you see in dim light.
Cones: photoreceptors that help you see in bright light, detecting colors and details.

Optic Nerve: the pathway between your photoreceptors in your eye and neurons in the brain that process visual information.

Visual cortex: the region of the brain with neurons specialized to process visual information like color, motion, shapes, objects, faces and more.

picture.

And that's how you perceive the world like a movie constantly streaming into your brain! Your eyes take in light, your photoreceptors detect it, and your brain turns it into the colorful and detailed world you see every day.

HEARING

If sight is all about light, then hearing is all vibration. When something vibrates, like a guitar string or your vocal cords, it creates waves in the air. Just like waves ripple on water, sound waves travel through the air in all directions, some of them coming in contact with your ears. Have you ever thought ears look funny? Well, their shape has a purpose! Your ears are like sound wave amplifiers, funneling sound waves into your eardrum like a net catches fish in water.

When sound waves enter your ear, those vibrations in the air make your eardrum vibrate too. Your eardrum is literally like a tiny drum in your ear, beating to the rhythm of incoming sound waves. This rhythm gets passed on to tiny little bones in your middle ear, called ossicles, which amplify the vibrations - basically turning up the volume on the music - and pass the rhythm onto your cochlea in your inner ear.

Your cochlea is a spiral tube filled with liquid deep inside your ear, and looks almost like a snail shell. And just like your eyes have "light-translators" - the photoreceptors that convert light into action potentials that neurons can understand - your ear has "sound translators." These sound translators are tiny hair cells in your cochlea, that turn the vibrations and rhythm of sound waves into action potentials - sending signals about sound to the brain in the language of neurons!

After your ear translates the rhythm of the world around you to the brain, your brain makes sense of those signals. Your brain is able to understand language, form music preferences and even help you create a map of the world so you know where different sounds are coming from. All from a simple vibration in the air!

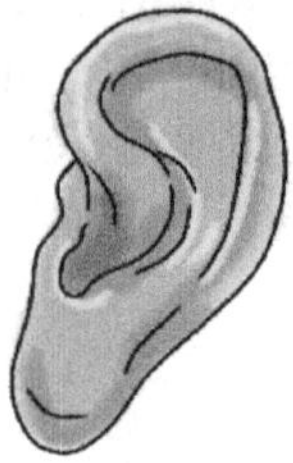

Key Terms

Sound waves: vibrations in the air.

Eardrum: also known as the tympanic membrane, a thin barrier between the outer and middle ear that transmits the vibration of sound waves.

Ossicles: tiny bones in the middle ear that transmit and amplify sound waves between the eardrum and cochlea.

Cochlea: a fluid-filled spiral organ in the inner ear that contains hair cells that convert sound waves into action potentials, sending neural signals about sound to the brain.

TOUCH

As with sight and sound, we need a new translator to convert signals from our skin to a message the brain can understand. To help this process, your skin is filled with special cells called touch receptors. These receptors convert information about pressure, texture, and temperature into action potentials to send to the brain. Sometimes those signals have a long way to travel. They can go as far from your finger or toe to your brain. That's why if you step on something sharp, or touch something hot, your foot or hand will react and quickly pull away seemingly before your brain has the chance to say "ouch!"

> **Key Terms**
>
> Touch receptors: special cells in your skin that convert information about pressure, texture, temperature, movement, and pain into action potentials, sending neural signals about touch to the brain.
>
> Somatosensory cortex: the region of the brain with neurons specialized to process touch information, with neurons dedicated to different parts of the body.

Your brain has a special region for processing these signals from your touch receptors called the somatosensory cortex. Located in your parietal lobe on the top of your head, the neurons in this region of the brain receive signals from specific touch receptors all over your body. Some neurons talk to the touch receptors in your hands. Others get signals from touch receptors in your feet. You have more neurons dedicated to processing information from your hands than your feet. That's why our hands are better at feeling details.

Your brain puts all the pieces of the puzzle together, making decisions about things that feel 'good' or 'bad' and integrates this information with other parts of the brain. Connecting with the visual cortex allows you to know which objects feel soft or rough, or connecting with the motor cortex allows you to move and respond to the world around you.

TASTE

You've probably already heard of your taste-translators without realizing what they actually do! Taste buds are specialized cells on your tongue that convert information about taste into action potentials, sending signals to your brain about different foods so it can decide what is yummy and what is icky.

Similar to how you have two kinds of light-translators (aka photoreceptors, rods for seeing in the dark and cones for seeing colors), you have five kinds of taste-translators (aka taste buds). You have taste buds for salt, sweet, sour, bitter, and savory. Taste buds specialize in these different tastes by interacting with different kinds of molecules in food.

Your brain then makes sense of these signals from your taste buds, integrating it with information from other senses and memories to form preferences and responses. Your mouth might even be watering right now reading this section because those brain regions connected to the experience of taste are telling you to expect food!

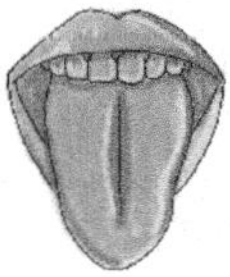

SMELL

Smell works very similarly to taste, and is also highly connected with the experience of taste. Your nose is also filled with specialized cells that convert chemicals in the air into action potentials, sending signals to the brain about smell. These smell-translators are called olfactory receptors, and they work the same way as taste buds. The signals they send to the brain are highly involved in our perception of taste, which is why smell is such a big part of appetite.

While you have five different types of taste buds, your nose contains about 400 different types of olfactory receptors! That is why smell often seems so much more complex than taste. Dogs are known for their great sense of smell, which makes sense because they have about 1,000 different types of olfactory receptors. More than twice as many as humans!

The brain processes these signals from olfactory receptors, connecting the experience to the other senses. Interestingly, the parts of the brain that process information about smell are closely connected to the parts of the brain that process memory, emotion, and motivation. Has a smell ever triggered an intense reaction? Triggered a sudden memory or emotion? Your smell-processing neurons are best buds with your memory neurons, creating these links in your brain and in your experience. Your brain stores a lot more information than what we can explicitly describe, sometimes much more than we are even aware of. What do your favorite smells remind you of?

Key Terms

Taste buds: specialized cells on your tongue that convert chemicals in food and drink into action potentials, sending signals to the brain about taste as salty, sweet, sour, bitter and savory.

Olfactory receptors: specialized cells in your nose that convert chemicals in the air into action potentials, sending signals to the brain about smell.

Did you know? You can store up to 1 quadrillion pieces of information in your brain during your lifetime.

Chapter 5: Learning & Memory

Strengthening connections

Learning and memory are a major function of your brain, allowing us to grow, change, bond with others, form opinions and preferences, tell stories and play music, have skills and hobbies, and experience the wonders of life. But how do we go from learning our A-B-C's to writing novels? From playing "Twinkle, Twinkle, Little Star" to Beethoven? From crawling to running marathons? From 2+2 to building rockets to the moon?

Our capacity to learn is immense, step-by-step storing more and more information. Learning and creating new memories involve complex processes constantly at work in the brain, involving many regions as we integrate senses with language and emotion with new experiences and memories.

Imagine your brain is like a library. When we learn something new, it's like adding a new book to the library. Sometimes we add a picture book, containing only images. Sometimes we add a manual, like how to ride a bike or throw a ball. The collection of shelves of books becomes our memory, and we can retrieve that information like retrieving a book from the shelf. Some memories may be displayed like new books on a big shelf right in front, while others may be old, dusty books, buried in the back or organized in the wrong section.

Let's take a closer look at how our brain writes, stores, and retrieves these books in our memory library:

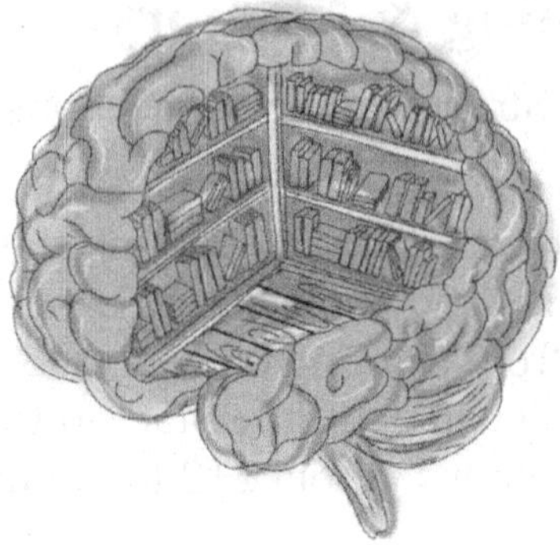

HOW WE LEARN

When we talk about learning in the brain, it involves much more than what you learn in school. Your brain is constantly learning, integrating new information from your senses, from people around you, from what you read, hear, see, and even from what you feel and do. When we learn something new or experience a new event, it triggers changes in the brain's neural connections. That's really what learning boils down to, changing the connections between neurons.

Remember that neurons talk to each other via action potentials - electrical signals that trigger the release of neurotransmitters. Neurotransmitters can carry all sorts of messages from neuron to neuron, and one neuron can talk to many other neurons, creating complex conversations with the language of the brain. Neurons will tend to keep having the same conversation if we keep giving them the same information, but a new song or a new flavor of candy triggers changes in the conversations.

These changes are known as synaptic plasticity, which is the ability of neurons to change their connections with each other throughout our lifetime. Your brain has the highest synaptic plasticity when you are young - your neurons are more flexible and willing to change, which is why people often say it's "easier" to learn new things when you are young.

One of the main ways neurons change their connections is by strengthening or weakening the existing connections between them. When one neuron constantly talks to another, the synapse between them will become stronger, like putting your best friend on speed-dial. This process is called long-term potentiation (LTP), which is the strengthening of synaptic connections following repeated stimulation. That is why "practice makes perfect" is a popular phrase, because the more you practice something new, the more you're allowing those neurons to talk to each other in that new way. If they talk enough, the connection becomes stronger and the "conversation" becomes easier.

Synaptic plasticity, or changes in the connections between neurons, can also take the form of creating brand new connections, or discarding existing connections, or altering the types of neurotransmitters or receptors within those connections. It's like the library of your brain can add chapters to books, combine similar books to make a series or a set, make edits or even throw away books. Your library is constantly being updated with each new experience!

TYPES OF MEMORY

Let's take a look at the kinds of books within the memory library of your brain. As our brain is constantly learning, creating new connections, and adding or altering the books of your library. The brain organizes these 'books' according to their type, importance, and relevance. We can think of these different types of memory as different sections in the library, with some closer to the front and easy to retrieve, and others secured in the back or even written on the walls themselves.

One section of your memory library is short-term memory, containing all

> **Key Terms**
>
> Synaptic plasticity: the brains ability to change by changing the connections between neurons.
>
> LTP: long-term potentiation, the strengthening of connections between neurons caused by repeated stimulation or "constant conversation."

the information you need for immediate use. It never keeps anything for long, and is like a live-stream assistant helping to accomplish the task at hand. It keeps information like what your friend just said as you're having a conversation, or directions from your teacher about how to play a game. Most of the time, this information gets thrown out as new information is constantly coming in. But sometimes this information is important, and gets transferred to another section: long-term memory.

Long-term memory keeps information from a few days up to an entire lifetime. Sometimes this means certain connections between neurons are so strong, they will always communicate. When many neurons establish these strong communication pathways, we call it a neural circuit. Neural circuits exist all throughout your brain, many connecting different brain regions to coordinate things like movement - made possible by connecting your visual cortex to your motor movements.

Memory can also be categorized by subject and type. Two main types of memory are procedural and declarative. Procedural memory is for procedures, things like riding a bike or playing an instrument. Declarative memory, on the other hand, is for facts and events. The knowledge you can "declare" or "describe." We can also organize declarative memory into two sub-categories: semantic memory (knowledge about the world) and episodic memory (personal experiences). Semantic memory is all the fun facts you know and contains many of the things you're tested on in school, whereas episodic memory works more like remembering episodes of a tv show - but a tv show of your life. These memories would be like retrieving the book of your favorite birthday party, or that family vacation, often filled with images and emotions to color the pages.

Overall, learning and memory are dynamic and complex processes that rely on the multitude of intricate interactions of neurons, synapses, neurotransmitters, and neural circuits within the brain. These processes allow us to acquire new information, retain experiences, and adapt to our ever-changing environment. What are some of the biggest sections in your memory library? Do you have a favorite book? What section are you adding to the most? Keep on learning and your library will become a vast world to explore.

Types of Memory

Short-term memory: short-term storage of information that you need for immediate use.

Long-term memory: long-term storage of information, from days to a lifetime.

Procedural memory: memory for skills and procedures, like riding a bike or playing an instrument.

Declarative memory: memory for facts and events, including semantic and episodic memories.

Semantic memory: knowledge about the world, facts, concepts, and ideas.
Episodic memory: memory of personal experiences.

Neural Circuits: a group of neurons that are strongly interconnected and communicate easily and regularly to perform a specific function.

Did you know? Your brain uses about 20% of your body's energy but only makes up about 2% of your body weight.

Chapter 6: Emotions & Personality

We don't often think of emotions as coming from the brain. We tend to describe the brain as rational and logical, a biological computer. Whereas emotions are felt in the body, and often seem separate from logic.

What makes you different from your brother? Your best friend? Are you shy, or outgoing? Are you adventurous? Funny?

BABY FACES

The human experience is emotional, a roller coaster ride of feelings from the moment we are born and enter the world screaming and crying. Do you think we born with emotions? Or do we learn them as we grow up?

Turns out, it's a bit of both. There are certain primary emotions that seem to be present from a very young age, and are recognized across different cultures from around the world with specific facial expressions. In fact, Pixar even made a movie about them! See if you can mimic these facial expressions and get someone to guess which emotion you are feeling.

Happiness, or joy, is often expressed with a smile. **Sadness**, feeling sorrow, disappointment, or grief, is expressed with a frown, and sometimes tears or a squinting of the outer edges of the eyes. **Anger**, feeling frustrated or irritated, is expressed with narrowed eyes, furrowed brow, and sometimes pursed lips or a grimace. **Fear**, feeling scared or in danger, is expressed with wide eyes and raised eyebrows, sort of like the opposite of anger. Similarly, **surprise** is expressed with wide eyes and raised eyebrows, often with an open mouth. Although surprise didn't make the Pixar cast, it's often counted as one of the primary emotions present from a young age and recognizable across cultures. And finally, as many babies trying new foods will express, **disgust** is shown with a wrinkled nose and a curled upper lip.

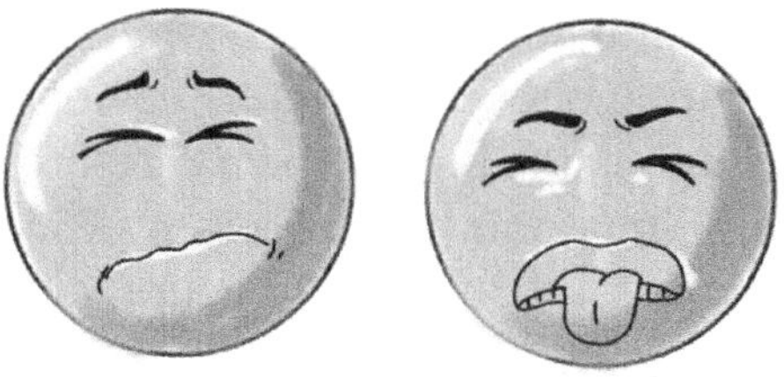

These primary emotions are all important for communication, survival, bonding, and learning. Some scientists say these emotions are innate - that is, present from birth - because of their evolutionary benefits. Fear and anger help people survive, avoiding or confronting threats. Surprise and disgust are also beneficial for survival, signaling potential danger or foods to be avoided, as well as providing social cues for acceptable behavior. Happiness and sadness are also important to understand in others in order to bond, communicate, and form relationships. Try having a conversation with someone or tell a story without making any facial expressions, and see how much more difficult it is to communicate!

COMPLICATIONS

As we grow older, emotions get more complicated and new ones come to join the party. We get introduced to things like jealousy, guilt, empathy, and gratitude. And next to stop by is pride, embarrassment, shame, and envy. And don't forget to leave the door open for contempt, hope, and regret. And many more that might not even have names, these emotions develop and express themselves in many different ways. So where do they all come from?

Emotions arise from a combination of neural circuits, influences from the environment, and a person's genes. As we navigate the world with the unique brain we were born with, we develop certain responses to the things happening around us. A hug might make you smile, and a dog might make you laugh.

As we learned earlier, neural circuits are pathways in the brain that allow for efficient communication between different regions. The amygdala, for example, is a brain region that is highly specialized for processing emotions. It's especially important for processing fear, and is quick to detect any potential threats. It then uses the neural circuits to alert other regions of the brain to respond to the threat, like telling your body to run away or your frontal cortex and occipital lobes to identify and remember the scary spider.

And just as certain brain regions are specialized in emotional processing, we have certain neurotransmitters - the messengers between neurons - that are important for emotions as well. Norepinephrine, for example, triggers those fight-or-flight responses to a threat. And serotonin is important for regulating your general mood, as someone who doesn't have enough serotonin might feel sad more often. And dopamine is important for motivation, and feeling good for being rewarded.

All these components in the brain and body work together, responding to the world around you. They help you navigate social communication, relationships, cultural norms, and behavioral dynamics. Emotions help us learn and remember important things, and give us context and help us develop intuition for new situations. And while we can trace certain emotions to particular brain regions or neurotransmitters, the experience of friendship or love is so much more complex than a neural circuit.

GETTING PERSONAL

While emotions are fleeting, personality is more stable. Emotions are often responses to the world around us, such as surprise for a birthday party or joy from a present. And while some can last a while, we wouldn't say they are permanent. Personality, on the other hand, is what we think of as our unique expression of humanity. The answer to "who are you?" or "what are they like?" When you think of yourself, how would you describe you? And where did any of that come from?

Personality is the unique expression of brain patterns, developed through a combination of your genes and experiences. Some might be more likely to be shy but had a very social childhood and grew up to be outgoing and talkative. Your personality is the unique pattern of thinking, feeling, and behaving, creating neural circuits in the brain to help you comfortably navigate the world. So, while these neural circuits can always change, they are more stable than the fleeting, daily emotions we experience.

Key Terms

Amygdala: brain region involved in processing emotions, identifying threats, and generating fear-based responses.

Prefrontal Cortex: brain region in the frontal lobe involved in reasoning, decision-making, regulating emotions and behavior.

Did you know? About 75% of your brain is made up of water, which is why drinking water is so important.

Chapter 7: Sleep & Health

Why do we sleep? What happens in the brain when we are sleeping? We spend about a third of our entire lives sleeping, so it must be pretty important. In fact, it is probably the most important factor in your general health! Think about slowly waking up after a long night of good, deep sleep. You feel cozy and warm, happy, and content. Now compare that to waking up to a loud alarm much too early after a restless night of tossing and turning, feeling grumpy and irritated at the world. It's no wonder the magic of sleep is so important for our health, both physical and mental.

Do you often remember your dreams, or forget them as soon as you wake up? Does it feel like your brain "turns on" when you wake up in the morning? While you were sleeping, your brain actually did a whole lot more than dreaming. Your brain also cleared away toxins that build up during the day, like brushing your teeth before bed. The memories you made and new things you learned get reinforced and consolidated in the brain, which is why it's so important to get a good night sleep before a big test. Sleep also helps restore energy levels throughout the brain and body, and promotes synaptic plasticity, your brain's ability to change and adapt.

THE SLEEPING BRAIN

Although it may feel like our brains "turn off" when we sleep, our brains are still active. However, the brain activity does look different from when we are awake. One way we measure brain activity is by measuring brain waves, which are electrical frequencies emitted from the brain. In Chapter 3, we talked about how neurons communicate with electrical signals called action potentials. We can measure these electrical signals, and these brain waves are the overall activity of the brain, or amount of neuronal communication in the brain. When we sleep, these brain waves change in predictable stages, and different frequencies are associated with different levels of consciousness. Let's take a look at these stages and what they feel like:

While we are awake, brain waves are fast. High frequencies (Beta & Gamma waves) are associated with intense concentration and focused attention for things like problem-solving and learning. As we start to relax, either by getting tired and "drifting off" or actively meditating to calm down, brain waves start to slow down. These brain waves are called alpha waves, and our brain emits them when we are still awake but in a very calm, relaxed state. Alpha waves are also associated with the "flow state" - a state of intense, focused attention on a task, or what some people call being "in the zone." You might see it with musicians that seem to be in their own world as they play, or an inspired scientist or mathematician solving a problem, or an artist in the process of creation.

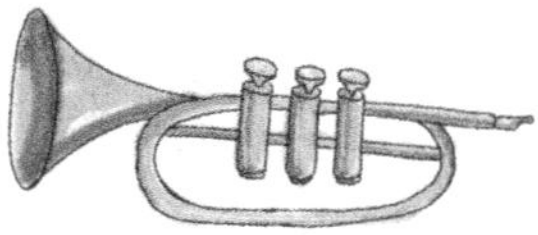

As we actually fall asleep, brain waves slow even more into Theta and then Delta waves. Theta waves are present during light sleep or daydreaming, and are associated with vivid imagery and creativity. Delta waves, the longest and slowest form of brain waves, are associated with deep, dreamless sleep. These frequencies of brain waves can also be observed in brains that are deeply meditating, and they are important for memory consolidation and clearing toxins from the brain.

Finally, and possibly the most important stage of sleep, is REM sleep. It is named rapid-eye-movement (REM) sleep because our eyes noticeably twitch under our closed eyelids during this stage. This stage of sleep seems to be the most crucial for memory consolidation, cognitive functioning, emotional processing, and general health. Interestingly, the brain waves emitted during REM sleep closely ressemble the fast, high frequency brain waves of being awake. This is also when dreaming happens, possibly explaining the random eye movements and higher neural activity. As far as your brain is concerned, dreams can seem as real as reality, while ignoring the laws of physics and logic! What wonders will you discover in your dreams?

HOW TO KEEP YOUR BRAIN HEALTHY

Sleep is the best way to keep your brain in tip-top shape, giving you the best chance of learning and feeling good. And the best way to ensure good sleep is to know how to take care of your circadian rhythms. Every animal has a circadian rhythm, which is a biological cycle regulating your body and behavior. Humans' circadian rhythms follow about a 24-hour cycle. This controls all sorts of important functions like your hormones, sleepiness, hunger, body temperature, energy levels, mood and immunity!

The purpose of the circadian rhythm is to maintain homeostasis, keeping internal stability and balance with a changing environment. It helps us keep our internal workings in sync with the outside world, allowing us to stay balanced and healthy. Our circadian rhythm takes cues from both within our body as well as from the world around us, and we can help keep this rhythm regular by knowing what those cues are.

Key Terms

Brain waves: electrical frequencies emitted from the brain, measured as a way to measure overall brain activity or amount of neuronal communication.

Consciousness: level of awareness

Memory consolidation: the formation and reinforcement of memories within the brain.

REM sleep: rapid-eye-movement sleep, characterized by high frequency brain waves and random movements of the eyes, when dreaming occurs, and important for memory consolidation and emotional processing.
Cognitive function: mental ability, including thinking, reasoning, learning, memory, problem-solving, decision-making and attention.

There is a small structure in the center of the brain called the hypothalamus, which is the main control center for your circadian rhythms. Like a conductor, the hypothalamus sends signals all throughout the brain and body about when to be hungry, or sleepy, when to warm up the body or cool it down, when to be social and energetic or when to be sleepy and moody. And the biggest influence on this master clock is light!

Light resets the hypothalamus, specifically a smaller structure within it called the suprachiasmatic nucleus (SCN), which is connected to the retina in your eye. Light is the main cue that regulates our circadian rhythm, telling our hypothalamus when to send signals for sleepiness or wakefulness, regulating our energy and hormones. The cells in your retina that communicate with the SCN are especially sensitive to blue light, which is naturally more abundant in the early morning. That is why looking at artificial bright lights late at night, like a phone screen or a TV screen, is so harmful to our health. But it also means that looking at bright light in the morning, especially natural sunlight, can help you wake up and easily reset your circadian rhythm.

Some other ways we can help regulate our circadian rhythms is by keeping a good schedule. This means going to bed and waking up at regular times. Whether you are a night owl or a morning lark, consistency is key. And keeping a healthy diet and doing some physical exercise during the day are also great ways to maintain healthy circadian rhythms, allowing you to sleep more easily and have more energy when you need it. So, limit that late-night light, eat well, and sweet dreams!

Key Terms

Circadian rhythm: biological cycle that regulates the body and behavior to maintain homeostasis.

Hormones: chemicals that regulate bodily functions (temperature, hunger, sleepiness, mood, etc.), carrying messages about what to do and when to do it. Similar to neurotransmitters, but found throughout the body instead of the brain.

Hypothalamus: control center for our circadian rhythms, regulating hormones and bodily functions.

SCN: suprachiasmatic nucleus, located in the hypothalamus, is the master clock of the brain, using light cues to regulate our circadian rhythms.

Did you know? You can have 4-7 dreams in one night, but you might not remember them!

Chapter 8: Brain Abnormalities

Just as your tummy can get sick with the flu, or your skin can come down with a rash, your brain can become ill like any other organ of your body. While an infection of your lungs can make it hard to breath, an illness of the brain can affect the way we think, behave, or perceive the world. Additionally, sometimes brains develop differently than most, and might be organized differently, resulting in patterns of thinking, behaving, and perceiving that are unique or rare.

A brain abnormality is anything that results in a brain that is different in some way from what scientists might call a "typical" or "normal" brain. Abnormalities may be structural or functional, and stem from genetics, illnesses, or accidents. But these terms are more about how common these brain differences are, rather than whether they are good or bad. Often, having a brain that works a little bit differently than the majority can be difficult simply because it is different, and many things in our society have been set up for the majority. When a brain abnormality results in difficulties in daily life, we call it a disability. That is why understanding these differences in brains is so important, so we can create a more welcoming world for brains of all types.

A DIFFERENT DEVELOPMENT

What psychiatrists call a mental disorder, or a mental illness, can also be thought of as a brain that works in a different way than the majority of brains. But while these differences may be the minority, they are still more common than we might think. Orange may not be a popular favorite color, but there are still lots of people that prefer a sunset hue to the blue sky or a pink flower. Over the years, scientists describe and categorize these differences as more and more people show similar differences in thinking, behaving, or perceiving the world, and come up with names and diagnoses. A benefit to categorizing differences in these ways allows scientists to better understand the causes, struggles, and potential treatments to ease any struggles that might arise from differences in thinking, behaving, or perceiving. However, that doesn't't make a diagnosis of a mental disorder necessarily bad, or even an "illness," as it is simply a slightly different brain.

For example, someone with a diagnosis of autism spectrum disorder might be hypersensitive to sensory stimuli. That means their brains "hear" sounds more strongly than most people, or bright lights will be much brighter to them. As we learned in Chapter 4, sensation is how we gather information from the world, but perception is how we really "feel" or understand that information, which happens in the brain. Sound enters the ear and light enters the eye, but some brains might interpret those same signals more strongly.

Someone with ADHD, attention-deficit/hyperactivity disorder, might have trouble paying attention in class, always seeming restless or fidgety. One explanation of these behaviors is that their brain may have less inhibition. For example, less of the neurotransmitters that say "shhhh" or "stop," or different pathways and neural circuits for brain regions involved in attention and motivation.

There are many behaviors and symptoms of brain abnormalities that scientists have categorized, which they are always updating to keep diagnoses accurate, personalized, and as helpful for treatment as possible. Talking about all of these diagnoses, their causes, and symptoms, would fill a book of its own. Here I simply want to convey that understanding how the brain works can increase our compassion and understanding for behavior that might at first glance seem odd or different. The more we can understand about the brain, the better we can adjust the world to fit all types of thinking, and reduce struggles that these differences might create.

Key Terms

Brain abnormality: anything that results in a brain that is different in some way from what scientists might call a "typical" or "normal" brain, either structurally, functionally, or chemically.

Mental disorder: a difference in someone's ability to think, feel, behave, or perceive that impairs, limits, or affects their daily life.

Diagnosis: an official label of a mental disorder, falling into a specific category and used to specify symptoms and treatments.

ACCIDENTS HAPPEN

A twisted ankle makes it difficult to walk, and a broken finger makes it difficult to write. Similarly, damage to the brain can cause difficulties for any of the many functions it performs.

Accidents happen, and sometimes our well-protected heads get hurt. When certain parts of the brain get damaged, it can cause damage to the functions or behaviors those brain regions are involved in.

For example, damage to the occipital lobes might result in impaired vision, while damage to the temporal lobes might cause difficulty hearing or understanding language. And damage to the frontal lobes might result in difficulty planning, reasoning, or properly thinking things through. Or damage to the amygdala, the emotional center of the brain, can cause an inability to regulate emotions or recognize danger. Even a hard bump on the head can temporarily impair brain function, which is why a concussion is no small injury. So, protect your precious brain and if you do anything risky, always wear a helmet!

DISEASES

Sometimes, brain damage can be caused by a disease, similar to how bronchitis damages the lungs or glaucoma damages the eyes. Diseases can affect the brain and cause changes in our ability to think, alter behavior, or limit perception. Depending on what areas, functions, or pathways a disease damages, it will affect different functions our brain is responsible for. Alzheimer's disease, for example, damages the brain's memory abilities, affecting behavior and thinking as well. Parkinson's disease, on the other hand, damages motor abilities by disrupting the brain's ability to coordinate movements. With the many wondrous things our brains can do, damage to any small part of the brain, or to any pathway, neurotransmitter, or function, can have serious effects on behavior. There are many diseases that affect the brain differently, though thankfully most remain relatively rare. And scientists and doctors are constantly working to understand and find solutions to the damage caused by brain-related diseases.

While some sources of diseases are known, many are not, and the best we can do is treat the symptoms. And the best way to treat these symptoms, as well as how to prevent or slow the disease entirely, is to understand how these diseases affect the brain. Additionally, the more we know about the brain and how it works when healthy, the better we are able to understand and treat the symptoms of an unhealthy brain.

Many scientists devote their lives to understanding the brain. Some focus on these kinds of common differences in the brain, whether they are causes by a disease, an accident, or a different development. Others focus on understanding the workings of a "typical" or a healthy brain, investigating the mechanisms of sight or memory or imagination. Both of these approaches provide insight to the spectrum of possibilities. Our brains are vastly complex, with much yet to be uncovered and understood, and each brain providing a unique perspective and ability. With your unique brain, what mystery might you solve?

Did you know? Each neuron in your brain can connect to thousands of other neurons, making a huge communication network.

Chapter 9: Consciousness

What is consciousness? Are you aware that you are aware? What do we mean when we say someone is un-conscious? Does consciousness reside in or come from the brain? Scientists, as well as philosophers, artists, spiritual practitioners, and religious leaders continue to debate about the nature and meaning of consciousness, as many humans have throughout time and space.

Imagine your mind as a movie theater where you're watching a movie. That movie is everything you see, hear, feel, and think. It’s both your sensation and perception, as well as your thoughts, memories, and emotions. Your consciousness is separate from all that, like the person sitting in the theater, watching the movie. It's the part of you that's aware of what's happening in the movie, aware of the world and the observer of the world.

Your consciousness is that ethereal “you” behind all of the fleeting experiences and activity of life, watching it all happen. It is the awareness of yourself and the world around you, of your unique thoughts, memories, feelings, sensations, perceptions, and environment. But also, the awareness comes from a separate layer, because you have the ability to recognize the observer. You are aware of your awareness. Does your brain hurt thinking about that? While we might be able to vaguely define and think about consciousness, we are still far from agreeing on where it comes from or how it works.

I THINK, THEREFORE I AM

The famous quote, “I think, therefore I am” was coined by French philosopher René Descartes in 1637. In a time of vast skepticism about the limits of human knowledge and reasoning, Descartes wanted to establish a foundation for knowledge that was beyond doubt. He realized that he could indeed doubt everything, a sensation and practice humans continue to do. As we have since learned, even our senses are mere electrical impulses translated from the energy of the surrounding world. However, he realized that he could not doubt the fact that he was doubting, recognizing that separate “me” behind the thinking and feeling and doubting. This famous quote has remained so popular because it simply points out that because we can think, and recognize that thinking, we must exist. We must be "real." Or at least, consciousness is "real."

CURRENT THEORIES

Some people say it's all an illusion, even consciousness. Everything is simply a consequence of biology and interacting matter, random exchanges of energy. Feelings, experiences, even pain aren't "real," only an illusion created in the mind due to the processes of evolution. These people would say that the universe is random and meaningless. Opponents to this viewpoint ask things like, "but why does a beautiful sunset feel so different from nails on a chalkboard?" There are no obvious evolutionary advantages to such intense differences in mundane experiences. Why would our brains create such vivid experiences *as well as* the ability to ponder them?

If we can agree that consciousness exists, does it come from the brain? As one scientist put, "What is it about the biophysics of a chunk of highly excitable brain matter that turns gray goo into the glorious surround sound and Technicolor that is the fabric of everyday experience?" Christof Koch (a computational neuroscientist, to be precise), who asked this question, is interested understanding where this technicolor comes from, and believes its basis might be neural activity. He also summed up this conundrum by saying, "the central puzzle of our existence: how a three-pound organ with the consistency of tofu exudes the feeling of life."

What makes our brain so special to have or create consciousness, whereas our stomach is left in the dark? Perhaps because it is where we integrate information? But there seems to be more to the story. Interestingly, while we can simulate sensory experiences by stimulating parts of the brain, we can't seem to simulate consciousness in the same way. People also report being retaining consciousness when brain activity has ceased (in near-death experiences), or having conscious awareness of different physical experiences (in meditation, out-of-body experiences, or lucid dreaming). In fact, meditation and lucid dreaming are skills that anyone can learn through training the mind in awareness, and offer interesting research avenues for understanding consciousness. And new insights into quantum physics may reveal secrets into the nature and depths of consciousness, as well as the universe itself. Are they reflections of one another?

WHAT WILL YOU DISCOVER?

Your brain is a complex and wondrous organ, but even more mysterious is the awareness of its own wonder! Humans have incredible abilities and capacity for growth, being able to share knowledge across time and space, investigate the unknown, discover the universe and create reality itself. Spend some time contemplating your consciousness, hanging out in that nowhere place of you-ness, and explore your own depths. You are more than the sum of experiences and neural connections, and you have the ability to watch them in peace. As you strengthen your consciousness, your ability to recognize the observer, the universe becomes a playground for discovery. What will you discover?

Did you know? Learning new things can help make your brain stronger and healthier as you grow.

Chapter 10: Brain Sciences

The Many Ways You Can Study the Brain

Are you interested in a career investigating the mysteries of the brain? Depending on how you imagine spending your days, and what you are interested in doing and learning, there are lots of possibilities when it comes to becoming a brain scientist. Are you fascinated by the workings of neurotransmitters, uncovering the biological mysteries of neuronal communication? Or are you more interested in the capacity of memory, the range of emotion, or the limits of attention? Do you want to discover and treat diseases, or create tools for enhancing perception?

Our brains are a magical mystery that we have only started to unlock, with many unanswered questions and unknown solutions to be discovered. Let's take a look at some of the most common ways people investigate the brain, what kinds of questions they seek to answer, the kind of education is required to get there, and how they spend their days.

A DAY IN THE LIFE OF A:

The first 3 careers we'll look at require a doctoral degree (Ph.D.), which focuses on the research aspect of understanding the brain. With a focus on research, the goal is to better understand the brain and grow our knowledge of the brain's workings and functions. Sometimes this is combined with clinical work, and these scientists also work with patients in developing treatments and therapies, but they don't diagnose patients, prescribe medication, or perform medical operations. The other kind of brain-investigating careers require a medical degree (M.D. or D.O), which focus on the diagnosis and treatment of conditions that affect the brain. With a focus on medical treatment, the goal is to better understand brain disorders and diseases in order to develop treatments and prevent symptoms. This involves a bit of research as well, but they spend more time with patients and focusing on treatments. Both are incredibly valuable and fascinating career paths, and have lots of variety within them!

NEUROSCIENTIST

Neuroscientists study the nervous system, including the brain and spinal cord (Central Nervous System), as well as its connections to the body in the Peripheral Nervous System. They investigate how the brain works, such as sensation and perception, learning and memory, or the functions of neurotransmitters. They do this by conducting research in laboratories, designing experiments, analyzing results, and publishing their findings. A Clinical Neuroscientist focuses on brain disorders and diseases, investigated their causes, symptoms, and inventing and testing treatments.

Education: Ph.D.

Typical Questions: How is smell related to memory in the brain? How do neurons communicate with each other? How do neurons change their connections? What causes neural plasticity? What causes conditions like Alzheimer's disease or schizophrenia, and how can we treat them?

PSYCHOLOGIST

Psychologists study human behavior and mental processes, using a variety of approaches and techniques including therapy, counseling, research, and assessments. Experimental Psychologists focus on research, investigating questions about how the brain works and understanding behavior, emotion, and cognition. They spend their days conducting experiments, analyzing results and publishing findings, teaching and collaborating. Clinical Psychologists, similar to Clinical Neuroscientists, focus on understanding and treating mental disorders, developing assessments, or categorizing behaviors into symptoms for diagnoses. They spend their days conducting therapy sessions, assessing the mental abilities of patients, conducting research, and often teaching as well.

Education: Ph.D.

Typical Questions: What are the effects of meditation on thinking and behavior? How do early experiences shape brain development and behavior? What behaviors are commonly seen in people with autism spectrum disorder? What reduces feelings of anxiety? How is sleep related to learning?

COGNITIVE SCIENTIST

Similar to a psychologist and neuroscientist, a cognitive scientist studies the brain using research and assessment. The main difference is that cognitive scientists are interested in learning about mental processes. This includes perception, memory, language. They take a more computational approach, utilizing a variety of fields including psychology and neuroscience, as well as philosophy, computer science, linguistics, and other fields of study.

Neuroscience focuses more on the biological aspects of the nervous system, and psychology tends to focus on the behavioral outcomes of mental processes, whereas cognitive science aims to understand the inner workings of the mind. They even develop computational models like artificial intelligence and machine learning to simulate these mental processes!

Education: Ph.D.

Typical Questions: How does our brain decide what to pay attention to? How does the brain encode and retrieve memories? Can computational models simulate the behavior of neural networks? What happens in the brain when we see a familiar face or hear our favorite song? What does brain activity look like during different levels of consciousness?

NEUROLOGIST

A neurologist is a brain doctor, someone who studies and treats diseases and disorders that affect the brain. Their days are spent evaluating patients, prescribing medications, performing procedures, and working with other healthcare professionals. They differ from Clinical Neuroscientists and Clinical Psychologists because they have a medical degree (M.D. or D.O.), rather than a doctoral degree (Ph.D.), and therefore specialize in the medical aspects of mental disorders rather than just the research and study of these disorders. They also differ from Psychiatrists in that they focus more on the biological or physical aspects of mental disorders, rather than the cognitive or behavioral effects of the disorder.

Education: M.D. or D.O.

Typical Questions: What are the underlying causes of Alzheimer's disease? What is the best technique for diagnosing traumatic brain injury? What signs or symptoms can we recognize before the onset of a neurological disease? What happens to the brain during a stroke? What regions of the brain are most affected by Parkinson's disease?

PSYCHIATRIST

Psychiatrists diagnose and treat mental disorders. They spend their days assessing patients, prescribing medication, providing therapy, or conducting research. Similar to a neurologist, but whereas a neurologist focuses more on the physical aspects of mental disorders, a psychiatrist focuses more on the cognitive and behavioral symptoms of disorders. They assess patients for cognitive ability, changes in behavior or differences in the way they think, feel, or perceive the world. They also provide and develop treatments, conduct research, teach, and collaborate with other healthcare professionals.

Education: M.D. or D.O.

Typical Questions: What is the best treatment for ADHD? How do early life experiences affect brain development? What mechanisms explain the benefits of meditation? What is the best treatment for depression? What risk factors are there for developing anxiety? How do medications alter brain functioning?

Key Terms

Ph.D.: doctoral degree, stands for "Doctor of Philosophy." Focuses on research and expanding the world's body of knowledge.

M.D. or D.O: medical degree, stands for "Doctor of Medicine" and "Doctor of Osteopathic Medicine." Focuses on diagnosing and treating patients.

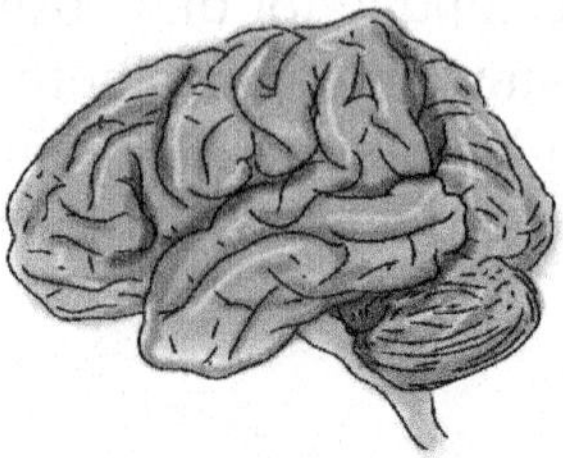

Made in the USA
Monee, IL
04 November 2026

67378734R00036